The Gulf

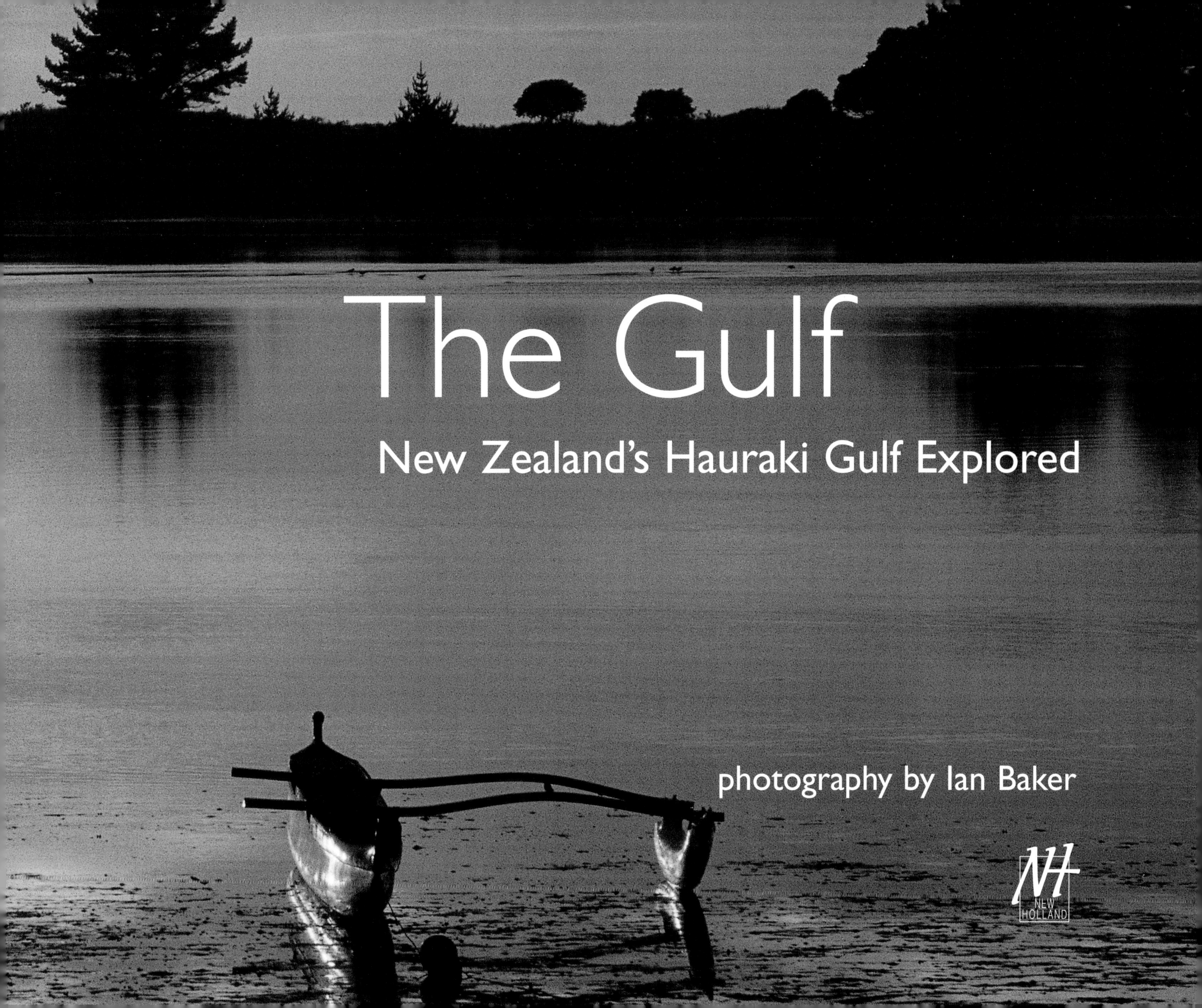

The Gulf

New Zealand's Hauraki Gulf Explored

photography by Ian Baker

NH NEW HOLLAND

This edition published in 2002 by New Holland Publishers (NZ) Ltd
Auckland • Sydney • London • Cape Town

218 Lake Road, Northcote, Auckland, New Zealand
14 Aquatic Drive, Frenchs Forest, NSW 2086, Australia
86–88 Edgware Road, London W2 2EA, United Kingdom
80 McKenzie Street, Cape Town 8001, South Africa

www.newhollandpublishers.com

First published in 1998 by C. J. Publishing

ISBN 1-877246-99-9

Publishing manager: Renée Lang
Contributing writer: John Morris
Design: Dexter Fry
Cover design: Gina Hochstein
Printed by Bookbuilders, Hong Kong

10 9 8 7 6 5 4 3 2 1

Front cover: Pleasure craft on the waters in front of Rangitoto Island.
Back cover: Te Kouma Harbour, Coromandel Peninsula.
Page 1: Tiritiri Matangi Island.
Pages 2-3: Ngunguru Bay, north-east of Whangarei.
Pages 4-5: Waiheke Island, looking across the Firth of Thames to the Coromandel Peninsula.
Pages 6-7: Miranda, Firth of Thames.
Pages 8-9: Duders Regional Park, looking north to the Gulf.

PHOTOGRAPHIC AND ILLUSTRATION CREDITS

ALEXANDER TURNBULL LIBRARY
Pages 12–13, page 14 (right), page 15, page 16, page 17 (right), page 89, page 146

AUCKLAND PUBLIC LIBRARY PHOTOGRAPHIC COLLECTION
Page 14 (left), page 17 (left), page 46 (right), page 124 (inset)

GARETH EYRES
Front cover

ROGER GRACE
Page 65, page 132 (inset): Poor Knights Islands, page 137

JOHN MORRIS
Page 52 (inset), page 53, pages 76–77

NEW ZEALAND HERALD
Page 154, pages 154–155, pages 158–159

PHOTOSPORT
Dean Treml: page 156, Andrew Cornaga: page 160

MIKE REEVES
Page 75

VISUAL IMPACT PICTURES
Pages 24–25, page 48

JAMES WAKELIN
Page 10: Map illustration

WINKELMAN COLLECTION
Page 46 (left), page 47

NAVIGATING THE HAURAKI GULF

AUCKLAND FERRY SERVICES

Fullers Auckland

ph 367 9111

Subritzky Line (car ferries)

Waiheke Island ph 534 5663

Great Barrier Island ph 373 4036

Water Tours

ph 376 1322

HELICOPTER CHARTERS

Heletranz

ph 415 3550

North Shore Helicopters

ph 424 4995

CHARTER VESSELS FOR HIRE

Auckland Boat Charters

ph 575 5922

Charter Link

ph 445 7114

Charter Services Auckland

ph 636 7541

Gulf Adventure Company

ph 444 2525

Sunsail

ph 378 7900

FISHING

Auckland Boat Charters

ph 575 5922

Sea Tours

ph 378 9088

AVAILABLE FOR FUNCTIONS

Chieftain Cruises Limited

ph 528 0052

Challenge Charters

ph 0800 500 171

Fullers Charters

ph 367 9115

Kawau Kat Cruises

ph (09) 425 8006

Function Cruises

ph 377 3877

Pride of Auckland

ph 373 4557

Sea Tours

ph 378 9088

SKIPPERED CRUISES

Auckland and Gulf Booking Agency

ph 303 0350

Gulfsail Charters

ph 536 5106

These companies are but a few of those listed in the Auckland Yellow Pages telephone directory (boat charter and hire section). Everything from diving trips to wedding receptions is catered for.

Contents

Hauraki Gulf
Rakitu
Mercury
GREAT BARRIER
Mokohinau
Kaikoura
Motuhaku
Poor Knights
Coromandel Coast
Little Barrier
Hen and Chickens
Firth
of Thames
Sail Rock
Bream Head
Pakatoa
Rotoroa
WAIHEKE
Ponui
KAWAU
Pakihi
The Noises
Moturekareka
Rakino
Mahurangi
Coast
Tiritiri
Matangi
Motuihe
Motutapu
Browns
RANGITOTO
N
AUCKLAND
Waitemata
Harbour

Introduction

One of the world's most splendid waterways, the Hauraki Gulf fans outwards from the Firth of Thames in the south, northwards to the open sea. The real Gulf is bordered on its eastern side by the dramatic beauty of the Coromandel Peninsula and the rugged grandeur of Great Barrier Island. The western borders are the sprawling suburbs of Auckland and the rural Mahurangi coast north to Bream Head. But the Hauraki Gulf Maritime Park, spilling out into the ocean to the north and north-east, extends this ultimate aquatic playground to over 7000 square kilometres.

Aucklanders have long enjoyed a love affair with the Hauraki Gulf. Not surprisingly, the 'City of Sails' claims more pleasure-craft per head than any other world city. An estimated 30 per cent of the population 'mess about' in boats of one sort or another. But the Gulf is much more than a paradise for boaties. It is a paradise for all leisure lovers. More than 40 dramatically different islands offer idyllic boating havens, pristine beaches, private picnic and camping spots, fascinating historical sites, superb diving, and wildlife sanctuaries of world importance. The Coromandel Peninsula is a mother lode of gold-rush history as well as a holidaymaker's delight. And necklacing the Gulf, all the way from the Firth of Thames through to the Whangaparaoa Peninsula and the sub-tropical charms of the Mahurangi coast in the north, is a magnificent string of ocean and inner harbour beaches.

'Hauraki' is Maori for north wind. Almost a thousand years after the first Polynesian canoes ranged the Gulf, warriors will again challenge the winds and waters of the Hauraki, this time in a quest for the holy grail of yachting — the America's Cup. An appropriate opening to the new millenium for a wondrous waterway that has no equal.

Chapter 1

Waters of History

(Pages 12-13) Sailing into history. Scows before the wind, 1907.
(Alexander Turnbull Library)
(Below) Captain James Cook.
(Auckland Public Library Collection)
(Below right) Hongi Hika.
(Alexander Turnbull Library)
(Far right) Auckland, the fledgling capital, 1843. Looking north-west towards Rangitoto Island.
(Alexander Turnbull Library)

Explorer's Delight

Rich in both Maori and Pakeha history, the Hauraki Gulf beckons and rewards the explorer. Reminders of a turbulent past — bloody wars, gold rushes, tragic shipwrecks — can be found in historic sites sprinkled across almost every island and along every stretch of coastline.

The first Gulf explorers were the Polynesian voyagers of around 1000 years ago. It is probable that even before the arrival of the European, the region centering on the Auckland isthmus — and including the Hauraki Gulf — had a population of over 20,000. Although constantly warring over land, there was a delicate balance of power between the Maori tribes. Then came the European and the musket. The pu — the gun — 'the great god of the white man' — became the ultimate weapon in intertribal disputes. Northland's Ngapuhi tribe acquired an impressive arsenal and, in 1821, swept south under the leadership of Hongi Hika. The island and coastal communities of the Hauraki Gulf were amongst the first to be laid waste. Today, relics of early Maori civilization survive on Browns (Motukorea), Motutapu, Little Barrier (Hauturu), Great Barrier (Aotea) and Waiheke Islands.

Captain James Cook, the greatest of the Pacific explorers, was the first European to sail the waters of the Gulf. In 1769, Cook observed the transit of Mercury at Whitianga on the ocean side of Coromandel. Rounding and naming Cape Colville, he sailed *Endeavour* up the Hauraki coast of the Coromandel Peninsula. He named these southernmost reaches of the Gulf the Firth of Thames. Frustrated by stiff south-westerly winds, Cook pushed no further into the Gulf than the islands of Ponui and Waiheke. With the wind still against him, Cook battled past the two northern guardians of the Gulf. He christened these islands Great Barrier and Little Barrier.

Cook's maps and reports kindled European interest in New Zealand and once the penal colony of Sydney was established in 1788, traders in search of timber and flax soon followed in the wake of the *Endeavour*. With the arrival of whalers, sealers and missionaries, the Bay of Islands became the centre of substantial general trade and the hub of an emerging British colony.

Following the signing of the Treaty of Waitangi in 1840, the seat of government was moved

CENTRAL
1897
A J ENTRICAN & Co Ltd
GENERAL MERCHANTS
JOHN
HOPKINS & COUTTS
BROWN BARRETT & CO
COFFEES & TEAS

(Far left) Downtown Auckland viewed from the North Shore, 1906.
(Alexander Turnbull Library)
(Above) Sir John Logan Campbell, known as the 'Father of Auckland'.
(Alexander Turnbull Library)
(Left) The Gulf's worst maritime tragedy occurred in 1894 when the luxury passenger ship Wairarapa, *returning from Sydney in appalling weather, slammed into the sheer rock face of Miners Head, Great Barrier Island. On that terrible night, 121 people lost their lives.*
(Auckland Public Library Collection)

south from Russell to Auckland. Beginning as a few huts and tents on the shores of the Waitemata Harbour, the new capital grew rapidly. European settlement also spread out across the Hauraki Gulf as farmers, miners and kauri loggers followed in the path of the whalers. By the 1850s Auckland was the most populous town in the most populous province in the country. Despite losing its status as capital to Wellington in 1865, Auckland continued to thrive. Today the region is home to 1.5 million people — nearly one third of New Zealand's population.

Auckland, the 'City of Sails', is the gateway to New Zealand for most international tourists. And the gateway to the Hauraki Gulf. The history of this cosmopolitan maritime city and the history of the unique waterway at her doorstep are inextricably intertwined. Exploring the Hauraki Gulf is a pleasure like no other. Ask any Aucklander.

Landmarks

(Far left) Gun emplacements, North Head. This volcanic headland, on the northern side of the entrance to the Waitemata Harbour, first served as a fort in 1885 during the Russian 'scare'. Visitors can still walk through turn-of-the-century defence tunnels. The Savage Memorial (top left) at Bastion Point (Takaparawhau) rises above the vault of Michael Joseph Savage, New Zealand's first Labour Prime Minister (1935-40). Overlooking the entrance to the Waitemata from the southern side, Bastion Point, also once a fort, is now a park. (Top right) Auckland's ferry building, completed in 1912 and completely refurbished in 1968, is one of the city's best known landmarks. (Below) The port of Auckland is New Zealand's largest and busiest, with shipping links to 160 ports in 73 countries. Another prominent landmark visible from the water is the Sky Tower — at 328 m, it is the tallest structure in the Southern Hemisphere.

Chapter 2

Living on the Edge

(Pages 20–21) Rooms with a view. Clifftop homes perched above the East Coast Bays on Auckland's North Shore.

(Above) Just enjoying time on the water is part of everyday life for some Herne Bay residents. Boathouses (opposite) are an attractive feature of the little bays that skirt the waterside inner city suburbs. In many cases, the owners need only amble down to the bottom of their road to gain access.

Under the Influence

The Maori named Auckland's eastern harbour Waitemata — 'sparkling waters'. One of the world's most beautiful and dramatic waterways, it opens into the Hauraki Gulf which stretches in a magnificent 250 kilometre sweep of coastline north and south of the city. From the Coromandel Peninsula in the south to the Mahurangi coast and Bream Head in the north, living, working and playing on the edge of this unique waterway is, for thousands, not only a way of life — it is the essence of life.

Soaring over the sparkling waters that separate downtown Auckland from the North Shore, the Harbour Bridge has been a city landmark since 1959. A more recent addition to the skyline is the Sky Tower (visible in the middle distance) completed in mid 1997. A popular attraction for tourists and Aucklanders alike, it boasts spectacular vistas across the city and Gulf.

70

One of Auckland's most prized residential locations, Takapuna Beach (far left, above and below) offers several kilometres of golden sand and a stunning outlook across to Rangitoto Island. Beachfront residences range from modest dwellings through to imposing mansions and, more recently, sophisticated apartment complexes.
South across the Waitemata Harbour lie its eastern suburbs counterparts — Orakei and Mission Bay — equally prized for their outlooks and easy access to the water.
Orakei (top left), boasts Paratai Drive, reputedly the most expensive street in the country, while Mission Bay (below and inset) is home to a thriving café culture.

More than a hint of the Mediterranean is apparent in the design of these eye-catching Gulf Harbour apartments on the Whangaparaoa Peninsula, less than an hour's drive from downtown Auckland. The specially designed canal in the foreground is an extension of the 1000-berth marina beyond, where 24-hour security and complete ship-fitting facilities make this complex the ideal haven for boaties. The town centre, featuring bars, restaurants and the exclusive Gulf Harbour Lodge, is just a stone's throw from the residential complex.

(Page 30) Truly a place to get away from it all, this cosy little holiday cottage, known as a bach in the local parlance, nestles on the shores of North Cove, Kawau Island.
However, time is running out for this Rangitoto Island bach (page 31). Once the lease is up the land will revert to the Department of Conservation.
A permanent home on the water is an option most people merely dream about, but it's a dream come true for the occupants of these houseboats (far left) moored near Thames.
A regular stopping-off point for the ferry that plies between the sleepy little settlement of Sandspit and Kawau Island, just 10 km offshore, Bon Accord Harbour (top) is a deep inlet on the western coast of the island.
Unoccupied for much of the year, these baches at Sandspit (below), 7 km east of Warkworth, come alive in the summer months when the local population is swelled by hundreds of holidaymakers who come to enjoy the good fishing or take a ferry ride over to Kawau.

The view is never the same from one day to the next from Auckland's eastern suburbs (main picture). And whatever the weather, there is always a stalwart craft or two out in the Gulf, overlooked by the brooding presence of Rangitoto Island, and supervised by a variety of seabirds (insets).

Wherever they live in the world, children are always drawn to play in gentle waves lapping at a sandy shore, and these children at Matapouri (far left), a popular holiday location north-east of Whangarei, are no exception.

Claiming a space on this crowded Devonport beach (above) is all part of the fun of living in New Zealand's biggest city.

But go north to Waipu Cove (below), a beach offering great surfing and fishing, south-east of Whangarei, and there is plenty of room for everyone— even in the summer months.

Secluded beaches and camping sites abound within a couple of hours' drive of Auckland. This idyllic spot is just around the corner from Long Bay, on the west coast of the Coromandel Peninsula.

Big Bay (below) is on the same coast, approximately 22 km further north.

UPSILON

Diners enjoy superb food and waterfront views at Cin Cin on Quay, one of Auckland's trendiest downtown brasseries. (Below) Auckland's Ferry Building, home to Cin Cin on Quay and the focal point for the commuter ferries which link the city with the North Shore and the islands of the Hauraki Gulf.

Time for a healthy fast food fix at Omaha (main picture), a seaside resort subdivision between the inner arm of the Whangateau Harbour and the ocean beach of Little Omaha Bay, 16 km north-east of Warkworth. Enjoying fresh fish and chips (inset) at a sports club on Great Barrier Island.

This unusual barbecue (below) has little in common with the modern gas-fired models of today — it started life as a float attached to the anti-submarine nets that stretched between Auckland's North Head and Rangitoto Island during World War II.

NACKS
CHEAP BAIT
PILCHARDS
$5
1KG BAG
MILK BREAD

Chapter 3

Boating Mecca

(Pages 44–45) Sunrise on the Mahurangi coast, an area rich in shipbuilding history. (Above) The famous centreboard 14ft yacht Rona, *scudding down the Waitemata Harbour in 1922.* Rona *was one of the early gaff rigged clinker built dinghies, which formed the popular X class fleet. The first of the design was owned by Lord Jellicoe and named* Iron Duke. *Thus the class was first called Rona/Jellicoe boats. (Winkelman Collection)*

(Right) Marine cavalcade. A Maori waka takes pride of place on the Waitemata Harbour, 1907. (Auckland Public Library Collection)

Success by Design

From the first Polynesian waka gliding across never-before-explored waters, to today's fast ferry surging towards the suburbs of Waiheke Island, packed with commuters, the marine pathways of the Hauraki Gulf have long been travelled by every imaginable type of craft. Sailing ships and schooners. Scows and steamships. Mullet boats and motor cruisers. Paddle steamers and power boats. Trawlers and tankers. Freighters and ferries. Container ships and catamarans. Warships and windsurfers. Cruise liners and charter boats. And of course, the yachts. Thousands upon thousands of sails whispering across waters that know no season. A dazzling kaleidoscope that encapsulates the 'City of Sails'.

An affinity with the ever-changing waters of the Waitemata Harbour and the Hauraki Gulf has always been both an inspiration and a challenge for boat designers and builders. As far back as 1841, the largest sailing ship to be built in Australasia, the *Sterlingshire*, was launched from the beach at Nagle's Cove on Great Barrier Island. And it is a tribute to the early boat builders that, in today's age of fibreglass, there are elegant 100-year-old kauri

(Left) The powerful gaff cutter Marangi *reaching down harbour in 1915. She was racing with the A class fleet to Mahurangi harbour, then as now, a popular destination. (Winkelman Collection)*
Over 80 years later, racing is heavily influenced by technology, epitomised by the America's Cup competitors such as the Italian entry sponsored by Prada (below), testing the waters for the 1999-2000 challenge, in which it was beaten by Team New Zealand. With the square rigged Soren Larsen *in the background, this picture encapsulates over 100 years of progress in sail.*

yachts still racing or cruising.

The 'City of Sails' enjoys a well-deserved reputation as a mecca of boat building. From the old flyers, fast cutters and sloops of the 19th century through to the super yachts and the multi-million dollar luxury motor yachts that are the Rolls Royces of the waterways in the 1990s, Auckland has long been a leader in the evolution of boat design.

The world's high regard for New Zealand boat builders was evident in the 1997-98 Whitbread round-the-world classic. Of the nine-strong fleet, two W60s (including Grant Dalton's *Merit Cup*) were built in New Zealand and six carried masts and rigging made here.

And then of course there is *Black Magic,* the elegant, powerful sailing machine that blitzed the America's Cup competition in San Diego in 1995 and brought the Auld Mug home to the 'City of Sails'. Team New Zealand went on to successfully defend the cup in 2000, the first country outside of the USA to do so.

AIR NEW ZEALAND

(Far left) Auckland is known as the City of Sails, but cruising and racing yachts share the waters with a huge variety of container ships, fishing boats and ferries. The transformation of the Viaduct Basin area on Auckland's waterfront has seen an influx of superyachts, attracted by the temperate climate and myriad sailing opportunities. (Above) Some sailors stay close to the shore but the Whitbread 60s (below), not much more than huge strong surfboards, challenge the world's oceans, sometimes racing at over 25 knots and surfing 10 m seas.

SEABOURN LEGEND

News

(Previous pages) Young admirers are favourably impressed by the sleek lines of the Seabourn Legend *(page 50). Auckland regularly plays host to a number of similar floating palaces, especially during events such as the America's Cup Challenge. (Page 51) America's Cup hopefuls test their skills at Maraetai. Auckland's maritime history has been well preserved. From the old triple expansion steam ferry* Kestrel *(right) later converted to diesel power, to the quaint little gaff rigged hard-chine dory* Delight *at the Mahurangi Regatta (below), all manner of historical craft still sail the Gulf waters.* The Spirit of New Zealand *Trust offers young people character building training voyages in sail (far right). The ship lies snugly at anchor at Okoromai Bay Heads near the international resort at Gulf Harbour.*

Chapter 4

Pleasure Playground

(Pages 54-55) Fishing in the shadow of Hen (Taranga) Island at Waipu Cove, on the western shores of the Gulf.
(Above) Kid kayakers test their skills on the waters off Motutapu Island, the site of a popular youth camp serving Auckland schools.
(Far right) Symbols of the 'City of Sails'. A windsurfer waits to catch a breeze in the lee of Rangitoto Island. In strong winds, speeds of 30 knots can be attained.

Endless Choices

It is a world apart. A world of sparkling waters, hazy, dreamlike panoramas and unique and fascinating islands. For sheer diversity of recreational opportunities, the Hauraki Gulf is unequalled. Endless waters, endless destinations. For the boatie there are a hundred favourite anchorages. A surfeit of choices is the only problem. Divers explore the underwater wonderland of the outer islands and the marine reserve at Goat Island near Leigh. Charter boat operators offer island alternatives — Rangitoto and Tiritiri Matangi for those who love to explore nature, Motuihe for picnickers, Waiheke for those with a taste for fine wines, arts and crafts and stylish cafés, Kawau for history buffs and Great Barrier for those who just want to get away from it all. And windsurfers, kayakers, water-skiers, anglers and beachgoers also have their favourite haunts. For the leisure lover, the Gulf's allure is irresistible.

An easy drive north of Auckland, the Mahurangi Regional Park (left) offers a welcome escape from the pressures of city life and is a favourite with anglers, bush walkers, picnickers, swimmers, kayakers and boaties. The Hauraki Gulf is paradise for an estimated 50,000-60,000 amateur anglers. The main catch is snapper, followed by kingfish. (Above) Time out in the heart of the city — hoping to catch the big one at Wynyard Wharf.

(Far left) Mission Bay, located in Auckland's wealthy eastern suburbs, is one of the city's most popular beaches. (Left) Baring it all at Oneroa, Waiheke Island. (Below) Kids relax at Mission Bay with a Project Jonah cradle and an inflatable whale. The cradle, which is attached to pontoons, is a New Zealand invention — now widely copied overseas — devised to refloat small stranded whales and dolphins.

(Far left) A jet skier rides the wake of a ferry. The Hauraki Gulf is one of the world's great aquatic playgrounds. (Left) Hunting for hidden treasure on Takapuna Beach. (Below) Sailing dinghies, Motutapu Island.

(Left and inset) A dolphin, divers and myriad marine life share the waters around the Poor Knights, at the outer reaches of the Hauraki Gulf Maritime Park. Rated by Jacques-Yves Cousteau as one of the world's top 10 dive spots, these islands are a mecca for thousands of underwater explorers. A marine and nature reserve, the Poor Knights protect many species that are rare or extinct on the mainland. A total ban on taking fish is strictly enforced. (Underwater photo: Roger Grace)

(Right) Dining out at Wenderholm Regional Park, a popular recreational area just north of Waiwera on the western shores of the Gulf.
(Below) Dining out at Mission Bay, one of Auckland's premier restaurant precincts.

(Above) Situated close to the densely populated areas of Pakuranga, Howick, and Bucklands Beach, The Half Moon Bay Marina shelters 500 boats. It is also home to the Bucklands Beach Yacht club – the biggest club in the southern hemisphere.

(Right) Gulf Harbour, stunningly located at the end of Whangaparaoa Peninsula, is the ideal base for cruising the Gulf. The marina services 1000 boats plus the craft owned by the canal dwellers living nearby.

Gates to the Gulf

From the early 1900s through to the early 1970s, a sheltered mooring was worth its weight in gold to the Auckland boatie. But sometimes, even on the best moorings, the back-breaking chore of rowing out to a bucking boat in a nasty chop had to be faced. Then came the marinas and the face of boating was changed forever.

Spanning a little less than 30 years, the marina industry has created a phenomenal interest in boating. Launches have doubled and tripled in numbers. Racing yachts have became cruiser/racers with all the luxuries of furling sails, pressure hot water, freezer/fridges, shower booths and private cabins. Sailing has become accessible to the 'less than die-hard' boatie.

Still, a love affair with a boat costs money. During the past 30 years of the marina and boat building boom, one maxim has held true: If you have to ask the price of a boat, you can't afford it.

(Pages 70-71) The sheltered sanctuary of Westhaven began as a mooring basin protected by a timber breakwater. Westhaven has always been the hub of Auckland's keeler racing, housing the Royal NZ Yacht Squadron, the Richmond Yacht Club and the Ponsonby Cruising Club. Moorings were gradually replaced with piles and walking access, and subsequently with the floating finger system of the modern marina. Westhaven now services over 2000 vessels and a charter fleet. Pile moorings are still much in evidence around Auckland's tidal estuaries and rivers. The pilings at Panmure (right) stretch up the Tamaki River to the Panmure Yacht Club.

Golf Gulf

A championship golf course is always appreciated, but a golf course that combines this with coastal cliff top holes can be considered a golfer's paradise. On the shores of the Hauraki Gulf can be found two of the world's finest marina based resort courses. The Gulf Harbour Country Club at Whangaparaoa and the Formosa Country Club at Beachlands were opened within weeks of each other and offer the golf enthusiast a visual golfing experience.

A tricky par four, the 12th is typical of the elevated greens at the Gulf Harbour Country Club, which offers an outstanding view of Auckland City and the Gulf. The elevated position also provides a grandstand view of the America's Cup regatta course.

(Above) The modern fully serviced clubhouse at Gulf Harbour. (Right) Teeing off on the 3rd hole with a backdrop of Gulf Harbour's stylish canal apartments, which adjoin the 1000 boat marina. (Below) The 15th hole (a par three), with its deceptive undulating green, begins the triangle of the three signature holes at Gulf Harbour Country Club. (Far right) In the foreground is the 15th hole. Beyond that the challenging 16th. At the far left of the picture is the 17th which runs adjacent to the cliff top properties at Okoromai Bay. (photo: Mike Reeves)

Million Dollar Holes

Located near the tip of the Whangaparaoa Peninsula, 40 minutes north of Auckland, the Gulf Harbour Country Club features a true championship golf course. Created by top international architect Robert Trent Jones Jnr, — who was given carte blanche and used the natural contours with exceptional flair — Gulf Harbour is considered to be as good as the best he has designed anywhere in the world. According to Jones, 'Architects rarely get to work with such magnificent topography anymore, which makes Gulf Harbour special.'

Adjacent to the stunning Gulf Harbour marina complex and canal apartments, the course uses land that was worth millions as building sites for some of the most spectacular holes in golf.

Jutting 15 kilometers out into the Hauraki Gulf, the Whangaparaoa Peninsula is exposed to the prevailing sou-wester and the occasional blow from the north-east. But all is forgiven when you play the signature holes — the 15th, 16th and 17th — situated on the cliff tops. They are three of the great golf holes of New Zealand, if not the world.

Being a resort course it offers more forgiving tees for the club golfer but from the back tees, at 6400 meters, low handicappers and professionals will find it extremely challenging.

Selected as the 1998 venue for the World Cup, Gulf Harbour is one of an exclusive group of clubs who have hosted this prestigious event. Sixty four of the world's best golfers competed on a course rated by the experts as among the game's top one hundred championship courses.

(Above) The stunning clubhouse at Formosa Country Club features a five-star restaurant and conference facilities. (Below) The signature hole, the 12th, a deceptively tough par 3 looking north-west to the Gulf, and (above right) the 18th hole with its man-made waterfall backdrop. The 16th hole (far right) is the scenic clifftop hole adjacent to the Pine Harbour Marina complex. Rangitoto Island, a Hauraki Gulf landmark, provides a striking backdrop to these coastal holes. (Photos: John Morris)

Big is Beautiful

Overlooking the glittering waters of the Hauraki Gulf from the fringes of the township of Beachlands, approximately 40 kilometres from Auckland's central business district, the Formosa Auckland Country Club is a triumph for its designer, New Zealand golfing great Bob Charles.

Covering 190 hectares, next to the Pine Harbour marina, the Formosa course occupies an area twice as large as the average golf club course. From the professional tees this is, at 6639 metres, the longest course in New Zealand. Generous use of the terrain has resulted in each hole being designed with its own unique character. Lakes cut into the fairway line and four coastal holes are among the most beautiful anywhere. Steep gullies along fairways and the front of several tees may be intimidating but they add to the excitement.

The huge, undulating greens are Formosa's most unusual and striking feature. Few courses in the world have greens as large, allowing options for difficult pin placements to challenge top class players or fairer pin placements for average players and visitors. Some of the greens are little 'domains' in their own right. This is the country club of the future. As with Gulf Harbour, Formosa is recognized by international visitors as offering golfing and cruising facilites that equal the best the world has to offer.

Chapter 5

Islands of the Gulf

(Pages 78-79) Afloat on the Gulf. The Junction Islands, Great Barrier. (Above) A brooding presence. Rangitoto — symbol of the 'City of Sails' and the Gulf. (Far right) Just a 30 minute ferry cruise from the bustle of the city, Rangitoto Island is the largest of Auckland's volcanoes. Rising 260 m above sea level, it towers over Browns Island, the tip of another, older volcano.

Worlds Apart

The Maori called them motu whakatere, 'the floating islands'. And, approached from sea level, they often do seem to be moving on the waters, merging, then drifting apart in the golden haze. There are more than 40 islands scattered across the Hauraki Gulf. Each has a distinctive character and its own unique appeal. From the spectacular volcanic lunarscape of Rangitoto to the golden sands and vineyards of Waiheke, the wild and rugged grandeur of Great Barrier to the colourful colonial history of Kawau, the internationally acclaimed wildlife sanctuaries of Little Barrier and Tiritiri Matangi to the world rated underwater wonders of the Poor Knights — each of the Gulf islands is a world apart.

(Above) Inhospitable to the extreme, the harsh lava rock of Rangitoto is, surprisingly, home to a distinctive plant community.

(Right) Rangitoto rises out of the dawn and over the approaches to the Waitemata Harbour. Maori described the island's distinctive skyline as the three knuckles of a local chief named Peretu — 'Nga Pona toru a Perutu'.

Symbol of a City

Rangitoto

It roared up from the depths around 600 years ago, the last and the largest of about 50 volcanoes that shaped the Auckland region. The Maori called it Rangitoto — 'bloody skies'. The youngest of the Hauraki Gulf islands, and the most striking, Rangitoto is to Aucklanders what the Eiffel Tower is to Parisians, the Statue of Liberty is to New Yorkers and Big Ben is to Londoners — the symbol of their city.

Possibly active as little as 300 years ago, Rangitoto is a peculiar place. There is almost no soil and no permanent water. The surface temperature can hit 70 degrees Celsius on a summer's day. Yet plants have adapted to this strange and alien environment and grow in unusual profusion. Rangitoto is home to over 200 species of native trees and flowering plants, 40 different ferns, several species of orchids and the largest remaining pohutukawa forest in New Zealand.

(Page 84) A city and its symbol. Auckland reaches towards Rangitoto Island.

(Page 85) The summit of Rangitoto offers unparalleled views of Auckland and the Hauraki Gulf.

(Left) Rangitoto has no permanent inhabitants, but day trippers can enjoy the salt water swimming pool and walk the hand-packed roads and tracks built by hundreds of prisoners during the 1920s and 1930s.

(Above) Exploring Rangitoto can be an adventure.

(Above) Islands in the stream — Motutapu, looking towards Rakino which was once owned by Sir George Grey. Twice Governor of New Zealand, Grey later bought Kawau Island and transformed it into his own personal Eden.

(Right) High times — Home Bay, Motutapu picnic, 1910. Picnickers listen to a speech by Prime Minister Sir Joseph Ward. (Alexander Turnbull Library)

(Inset) Quieter times — Home Bay 1998.

Pa and Picnics

Motutapu

Tied to Rangitoto by a natural causeway, the sacred island of Motutapu was once the site of intensive settlement by both Archaic and Classic Maori. Enriched by the volcanic fallout from Rangitoto's 'bloody skies', the island is now productive farmland. It is also archaeologically important. Well over 300 historic Maori sites have been recorded and even today, evidence of their occupation can clearly be seen in the terracing which covers much of Motutapu. Remnants of old Maori forts — pa — share the island with fortifications built to protect Auckland during World War 11.

Motutapu was phenomenally popular with Victorian picnickers. According to the *New Zealand Graphic*, 'No distinguished visitor that comes to Auckland, but makes a trip to Motutapu'. It was not unusual for hundreds of picnickers to be ferried over to Home Bay for the day. A major Motutapu picnic could attract as many as 5000 merrymakers.

(Left) Enriched by the ash of Rangitoto's eruptions, most of Motutapu is a working farm. European settlers began farming the island in 1840. (Above) Home Bay, a favourite destination for Auckland's leisure lovers.

Lying between Rangitoto and the suburbs of Auckland, Browns Island is the tip of a volcano that erupted around 20,000 years ago. Originally known as Motukorea, its grassy slopes are still marked by the walls and trenches of Maori settlement.

John Logan Campbell — the 'Father of Auckland' — and his partner William Brown purchased the island from local Maori in 1840. Farmed for almost 150 years, Browns became a public park in the 1950s when philanthropist Sir Ernest Davis bought the island and presented it to the people of Auckland. Access is limited to private boats as there is no public ferry service to the island.

Musick Point (foreground) is named after Captain Edwin Musick, who piloted the first Pan American flying boat from San Fransisco to Auckland in March 1937. An art deco building erected as a memorial was used for a time as a radio station.

(Above) Snapper Bay, an anchorage favoured by day tripping picnickers. (Right) Motuihe's tranquil appearance belies a colourful history — the island was a quarantine station for Auckland, a World War 1 POW camp and a naval training base.

Island Escape

Motuihe

Now the most popular picnic destination in the Gulf, Motuihe, too, was once densely populated by Maori. But the island's most famous resident was the 'Sea Wolf' — Count Felix Von Luckner, the buccaneering pride of the Kaiser's Imperial Navy. Confined to Motuihe after his capture in World War 1, Von Luckner, together with other German prisoners, pulled off an elaborate and daring escape in 1917. Dozens of Auckland yachties gave pursuit to Von Luckner's commandeered timber scow, playing hide and seek around the Gulf and down the Coromandel coast. Von Luckner was finally recaptured in the Kermadec Islands, nearly 1000 kilometres north of Auckland.

(Far left) Woody Bay, Rakino. Privately owned, the island has around 25 permanent residents and is popular with holiday-makers. Rakino is also ideally situated to offer grandstand views of the America's Cup races. (Above) Looking across Rakino to the Noises Islands. (Left) Rakino's Home Bay jetty awaits the holiday rush. (Right) Faded but functional — the island's telephone box needs a coat of paint.

Suburbs on the Sea

Waiheke

(Pages 98–99) The Noises were likened by a French sea captain to scattered nuts, or 'noisettes', a moniker which he charted but which was subsequently 'anglicised' to Noises. (Above) The 'knuckles' of Rangitoto viewed from the southern end of Waiheke. (Right) Matiatia Bay, gateway to Waiheke. Auckland sprawls in the background, a mere 35 minutes away by fast ferry.

Originally named Te Motuarairoa, 'the long sheltering island', Waiheke was subjected to successive waves of Maori tribal occupation for around a thousand years. The Gulf's second largest island also held appeal for the early European explorers.

'I particularly noticed on Wai-Heke sites that would be admirably fitted for settlement,' declared the French navigator Dumont d'Urville in 1827. A prophetic observation. Today Waiheke is New Zealand's third most populous island.

Until about 10 years ago, Waiheke was known mainly for its population of eccentrics, hippies, alternative lifestylers and artists. Then the slow ferries were replaced by Fullers' Quickcat. The 17 kilometre trip from downtown Auckland was cut from an average 75-90 minutes to just over half an hour. Waiheke was 'discovered' and is now virtually a dormitory suburb with around 7000 permanent residents, including 1200 daily commuters. During the Christmas holidays, the population can swell to 30,000.

But the island's casual, laid back atmosphere has survived the commuter invasion. Waiheke retains its reputation as a haven for artists with more than 70 resident writers, painters and craftspeople. The island has also gained an international reputation for the art of boutique winemaking. Boasting more than 25 vineyards, Waiheke produces some of the country's best red wines. In 1997 the island claimed a world record price for a New Zealand wine when a large bottle of Goldwater Waiheke Island Cabernet Merlot 1990 fetched NZ$6000 at a New York auction.

Despite an ever increasing population, attracted by the more relaxed lifestyle and a climate which is generally warmer than Auckland, with less humidity and rain and more sunshine, Waiheke still retains its 'alternative' appeal.

The carparks at Matiatia Bay are full and the commuter ferry prepares to leave for Auckland. In addition to the passenger service to downtown Auckland, a vehicle ferry, which also takes foot passengers, makes several daily sailings from Half Moon Bay in the city's eastern suburbs. A passenger ferry between Auckland and Coromandel also makes regular stops at Waiheke.

A vintage view. The Peninsula Estate vineyard, dramatically positioned on Hakaimango Point, is one of 25 Waiheke wineries. The island's ideal growing conditions — hot, dry summers and stony soils — have produced some of New Zealand's best red wines. Olives — another crop suited to the Mediterranean style climate — are also becoming a Waiheke speciality, with orchards of up to 3000 trees being planted. Wine tours are one of the island's major attractions. (Below) Crushed grapes at Putiki Bay vineyard — one of Waiheke's newest wineries and the first to make Pinot Noir.

(Left) Established by owner Stephen White in 1982, the Stonyridge Vineyard is Waiheke's most prestigious winery. Lying on stony but free draining soil, just one kilometre from the sea, it is the first New Zealand vineyard to have all five Bordeaux varieties: Cabernet Sauvignon, Cabernet Franc, Merlot, Malbec and Petit Verdot.
(Below) The Stonyridge winery.

(Right) Despite the influx of commuters, Oneroa, Waiheke's main shopping centre, retains its charming seaside resort atmosphere. The island's eateries range from vineyard and a la carte restaurants to wine bar cafés and family cafés. Waiheke's culinary scene is laid-back — like everything else on the island. (Below) Relaxing Mediterranean style — the Mudbrick Vineyard and Restaurant. (Far right) Suburbs on the sea. The appeal of Oneroa is obvious. In a few short years, real estate prices have risen almost 100 per cent. Waiheke has become a magnet for those seeking a better or different lifestyle and the influx of escapees from Auckland shows no sign of waning.

(Far left) North Harbour, Ponui Island, is a favoured anchorage for yachties cruising 'the bottom end' — the colloquial term for the waters around the eastern shores of Waiheke. (Left) Rotoroa, a near neighbour of Ponui, is the site of a Salvation Army administered hospital for chronic alcoholics. Landing is prohibited. (Below) Landing is certainly welcomed at the nearby private resort island of Pakatoa.

Tiritiri Matangi — 'looking to the wind' or 'tossing in the wind' — is an open wildlife sanctuary, unlike those on some other islands. Here, the public can view rare animal species in the wild. Farmed continuously from the mid 1850s until the 1970s, the island is now being allowed to revert to its natural state. Tiritiri's 20.5 m lighthouse has been a beacon to ships on the Hauraki Gulf since 1865. (Inset) The takahe (top) and the saddleback (below) are two of the rarer birds introduced to Tiritiri by the Wildlife Service and the Department of Conservation.

Remnants of history. Lying in the shallows at Motuerekareka, one of the many tiny islands south of Kawau, the steel barque Rewa *was a complete ship when she was beached in 1931 to form a breakwater. Originally named the* Alice B. Leigh, *she was, at 3000 tons, the biggest four masted barque afloat in her day. (Below) For some, the* Rewa *is an excitingly different 'adventure playground'.*

Colonial Eden

Kawau

Once a Maori stronghold, Kawau was uninhabited when Governor Sir George Grey purchased it for £3700 in 1862 and set about creating his own private Eden. The local tribe had been decimated earlier in the century by the muskets of the Ngapuhi, and the manganese and copper mining, which had brought Europeans to the island in 1838, ceased in the early 1860s. Kawau was left to a man with a dream.

Moving into the mine manager's house, Grey remodelled and enlarged it. No expense was spared. Eventually he would lavish close to £100,000 of his personal fortune on the estate.

Up to 17 gardeners tended exotic plants imported from all over the world. In 1876 a visitor noted that Grey had a collection of trees and shrubs that 'almost defied description'. Mixing with the native pohutukawa, rata and kauri were Brazilian palms, Indian rhododendrons, blue jacarandas and red gums, wattles and eucalypts from Australia, silver firs from South Africa, cork trees, walnuts, olives and oleanders from the Mediterranean, English oaks and elms and Fijian climbing plants and spider lilies. The catalogue was 'almost endless'.

Mansion House, an impressive reminder of Hauraki Gulf history, is open to the public.

Grey introduced zoological specimens with lavish disregard of expense and suitability to the environment. Peacocks paraded, kookaburras screeched. Chinese pheasants, Californian quail, Cape Barren geese, wild ducks, turkeys, guinea fowl and emus wandered at large. New Guinea tree kangaroos hopped up amongst the puriri branches. From Britain came deer; from South Africa antelope, monkeys and zebras; and from Australia kangaroos and wallabies.

The ecological impact would prove disastrous — but guests of the day were overwhelmed. A friend of Grey's, British historian James Froude, thought his estate was 'as pretty as Adam's garden before the Fall'. In 1888 Eden did fall. Age and ill health forced Grey to sell up and leave. His mansion became a boarding house, then, for a period, a licensed hotel. Restored to their original splendour in the 1970s, Grey's Mansion House and surrounding grounds are elegant reminders of the Hauraki Gulf's rich and colourful colonial history.

(Far left) Historic Mansion House Bay, a tourist magnet and a favourite anchorage for Auckland yachties. (Top left) Held over three days, the Furino fishing tournament attracts up to 3000 anglers, all hoping to hook the big snapper — and a share of the $100,000 prize list. (Below) A prize 13.02 kg catch — the winner of the 13th Snapper World Cup.

A Kawau landmark. The ruined copper mine engine house is all that remains of a once profitable industry that at its peak, in the mid 1840s, employed over 200 people. (Far left inset) The ore-stained entrance to a long abandoned coppermine. Some shafts ran under the sea — one to a depth of 24 fathoms. Engineering problems and flooding — together with the lure to the miners of the Otago and Australian goldfields — led to the closure of the mines in the early 1860s.

(Below) Squadron Bay, Bon Accord Harbour — home to the Kawau Yacht Club.

North Cove, a prize gem in the necklace of Kawau anchorages, is also a port of call for the local mail ferry which services 63 wharves and jetties around the island. It is claimed to be the largest water mail run in the southern hemisphere. (Below) Schoolhouse Bay, another port of call on the mail run. The original schoolhouse — since modified several times — was built by Sir George Grey who brought his niece out from England to teach the children of miners.

Gulf Guardian

Great Barrier

Rugged and grand, Great Barrier is the Hauraki Gulf's largest island. First settled by Maori about 1000 years ago and named Aotea — 'white cloud'— this remnant of the Coromandel Peninsula has been severed from the mainland for 10,000 years. Lying 90 kilometres north-east of Auckland, it guards the outer reaches of the Gulf from the excesses of the Pacific.

Great Barrier has a hard pioneering history marked by ruthless exploitation. The felling of the tall stands of kauri that once distinguished the island began on the shores of Port Fitzroy around 1794. The plunder of this natural treasure — referred to as 'that great phenomenon of Botany' by the naturalist Ernst Dieffenbach — was virtually completed before the end of the 19th century.

The discovery of gold at White Cliffs (Te Ahumata) in 1892 saw seven different companies working claims and the population of Oroville swelled to 500. By 1910 the rush was over and today little can be seen of the town. After the miners came kauri gum diggers and then settlers who laid waste to the forests with fire to establish farms. And worse followed. In 1925, the Kauri Timber Company decimated the remaining kauri, milling every tree 30 cm or more thick. A mere 40 hectares survived around the rugged peak of Mt Hobson (Hirakimata). In the second half of this century the New Zealand Forest Service began restoration of the island's natural cover and today over 60 per cent of this wild and beautiful place is protected as conservation estates or reserves.

Although readily accessible by fast ferry or aircraft, New Zealand's fourth largest island retains a splendid sense of isolation. With a permanent population of just 800 individualists who practise conservation and alternative lifestyles, Great Barrier has long been a favoured 'get away from it all' destination for Aucklanders.

(Far left) Little Barrier Island — known in Maori as Hauturu, 'the resting place of the winds', is one of the world's most important wildlife sanctuaries, and landing without official permission is strictly forbidden. (Above) Wild and beautiful, Great Barrier retains its sense of isolation.

(Left) All that remains of the Kauri Timber Company, one of the most ruthless exploiters of Great Barrier's resources. The company was responsible for taking 55 million feet of logs off the island, before ceasing operations in 1941. (Inset left) The Kauri Timber Company, 1910. Their sawmill at Whangaparapara Bay soon became the largest producer in the southern hemisphere. (Auckland Public Library Collection) (Below) The ruins of New Zealand's last shore based whaling station, which closed in 1961, can also be found at Whangaparapara. In one year alone they killed more than 100 humpback whales off Great Barrier Island.

(Far left) A Tryphena landmark. The original Blackwell family cottage, built in 1880. Descendents of this well-known pioneering family still live in the area. (Left) Nikau palms, symbols of Great Barrier's unspoiled 'get away from it all' appeal.

(Below) Great Barrier's main peak, Hirakimata (Mount Hobson).

Freedom is a way of life on Great Barrier – horseriding on Medlands Beach. Although only a two hour cruise or a 40 minute flight from Auckland, 'the Barrier' remains a world of its own.

In 1922, the 12,000 ton steamship Wiltshire, *on a run from Liverpool to Auckland, was wrecked near Rosalie Bay, on the southern coast of Great Barrier. Her anchor now rests at Whangaparapara.*

(Right) A reminder of how inhospitable the island's shores can be.

(Below) A tobacco tin washed ashore from the Wiltshire.

(Far right) Hauraki Gulf history left high and dry. During World War 11, these floats supported anti-submarine nets stretched between Auckland's North Head and Rangitoto Island.

Lapped by the outer waters of the Gulf, the volcano sculpted Poor Knights (far left) attract thousands of divers, Once a natural fortress for Maori, these islands are now an absolute marine and nature reserve.
(Inset) Clear, warm currents attract an extraordinary array of marine life.
(Underwater photo: Roger Grace)
(Left) Waipu Cove, on the western shores of the Gulf, looks out to the Hen and Chicken Islands. Once occupied by Maori, the group is now a protected marine and wildlife sanctuary, and is second only to the Poor Knights as a popular diving spot.

Character Coast

The beaches of the Hauraki Gulf range from long stretches of sweeping golden sands to intimate coves. (Above) Looking north towards Sandy Bay. (Right) Ngunguru Bay, north-east of Whangarei.

South of the Whangaparaoa Peninsula, the western waters of the Hauraki Gulf lap the shores of suburbia. To the north of the peninsula, the coastline forms a sub-tropical mosaic of rocky inlets and peninsulas, peaceful, isolated coves and rolling surf beaches.

Governor William Hobson at one time considered the Mahurangi a likely location for the country's capital but, in 1840, decided to establish Auckland on the shores of the Waitemata. Today, the Mahurangi coast is increasingly attracting those seeking a more relaxed lifestyle but for the original European arrivals, timber was the lure. Long before the main wave of immigration, which brought farmers and boat builders to the region in the 1850s, lone traders settled along the shores of the Gulf and bartered with local Maori for kauri and rimu.

In the late 1820s, as the coastal forests were cut out, the timber men moved up the winding Mahurangi River and a small community sprang up on the site of Warkworth. For many years the only link between Warkworth and Auckland was by ship. The intense rivalry between the two companies providing regular services to Auckland culminated in 1905 with the *Claymore* running down and sinking the *Kapanui* in Waitemata Harbour.

Over the years, the region was to attract many disparate groups of immigrants. Amongst the most prominent were the Scottish Highlanders who, led by the fearsome Reverend Norman McLeod, founded Waipu. When McLeod died it was felt no minister could replace him. It is claimed that his pulpit was dismantled and the pieces shared out amongst his followers. Today descendants of Waipu's founders number tens of thousands throughout the country.

The Bohemians who settled Puhoi also left a legacy. The New Zealand expression 'up the boo-ay', meaning lost in the wilderness, allegedly derives from 'up at Puhoi' which was one of the most isolated communities in the 1860s. Today it is less than an hour's drive from Auckland.

(Right) Pakiri Beach, just north of Leigh, is famous for its white silicon sands and is favoured by surfers and horsetrekkers. (Below) The sheltered waters of Leigh Harbour provide an excellent base for fishing boats working the Gulf. (Far right) Feeding time for a shoal of blue maomao and parore at the immensely popular 'Goat Island Marine Reserve' as it is known to the nearby Leigh residents. Established in 1975 as New Zealand's first marine reserve, this underwater wonderland is officially known as the Cape Rodney to Okakari Point Marine Reserve. (Underwater photo: Roger Grace)

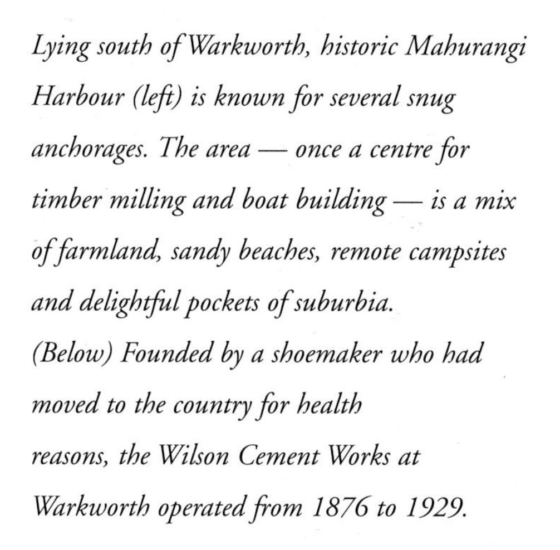

Lying south of Warkworth, historic Mahurangi Harbour (left) is known for several snug anchorages. The area — once a centre for timber milling and boat building — is a mix of farmland, sandy beaches, remote campsites and delightful pockets of suburbia.

(Below) Founded by a shoemaker who had moved to the country for health reasons, the Wilson Cement Works at Warkworth operated from 1876 to 1929.

In 1863, when the first Bohemian settlers were paddled up the Puhoi River (right) and dropped at the landing now known as Puhoi Village, the area was wild and isolated. Today, it is an easy drive from Auckland. Puhoi means 'slow water' in Maori and refers to the gentle tide that ebbs and flows to the open sea at Wenderholm, 8 km down river. (Above) Puhoi's Church of St Peter and St Paul dates from 1881.

Just north of Waiwera, Wenderholm Regional Park (right) includes the mouth of the Puhoi River and surrounding coastal forests. With a safe sandy beach, walking tracks through magnificent native and exotic bush, numerous picnic spots and historic Couldrey House which was once a base for visiting royalty, Wenderholm is a favoured destination with day-trippers.

(Far right) Sculpted by nature — pohutukawas at Wenderholm.

Coromandel Fortunes

(Above) Miranda, on the south-western shores of the Firth of Thames, is a mecca for migratory birds. More than 12 species come here each spring.
(Right) The most spectacular displays of pohutukawa — a Gulf emblem — are found along the Coromandel coast. Known as the New Zealand Christmas Tree, because it flares into crimson display from late October and into the Christmas holiday period, the broad canopy of the pohutukawa can grow as high as 20 m.

Jutting claw-like out into the ocean, the Coromandel Peninsula forms a dramatic protective barrier along the eastern reaches of the Hauraki Gulf. Sharing many of the characteristics of the Gulf islands — the ever-present sea, a narrow mountain backbone and plant life that has more similarity to that on the neighbouring islands than to other mainland regions — the peninsula also shares a similar history of exploitation.

Kauri timber and gum and gold drew men to the Coromandel from all over the world. The first timber-seeking British ship sailed up the Firth of Thames in 1794, just five years after Captain James Cook explored these waters. By 1820, when HMS *Coromandel* anchored in a harbour near the top of the east coast to take on kauri spars — and leave her name on the harbour, the nearby township, the volcanic mountain range and the peninsula itself — the plunder was at its height. During the 19th century millions of cubic feet of kauri were taken. By the 1930s logging had faded away. Only remnants of the once magnificent kauri forests remain today.

Although profits from the timber trade were enormous, the peninsula's most valuable resource remained hidden until 1852 when New Zealand's first gold rush took place just north of the present township of Coromandel. It was short lived when the miners discovered it was not alluvial gold, but embedded in quartz, and had to be hard won with pick and shovel.

The real rush for riches began in 1867 when gold was discovered in dazzling quantities at Thames. The first reports were so extravagant that few believed them. But as an observer of the day recalls, 'When solid bars were exhibited in a jeweller's shop in Queen Street, citizens took the fever ...' By 1870 the boom town's population was around 20,000, greater than that of Auckland at the time. Thames' goldmines produced ore worth £7,000,000. New fields were discovered at Coromandel, Kuaotunu and Karangahake. The Martha mine at Waihi — one of the richest in the world — opened in 1892 and delivered gold to the value of £28,469,000 until its closure in 1952. Continued mining on the Coromandel is a controversial subject.

Exploited since the earliest days of European settlement, the Coromandel Peninsula remains serenely beautiful. Volcanic plugs rise above the rainforest, creating the unique landscape that is the Coromandel Ranges. The picturesque western shores offer views across the shimmering Gulf to Auckland. The white sand east coast beaches are some of the best in the country. And echoes of the frenetic gold rushes that swept over the region can still be found in many of the towns. The Coromandel Peninsula retains its riches.

SCHOOL OF MINES
1898

(Far left) Gold-rush fever — 19th century miners. At one time, there were an estimated 700 productive mines in the vicinity of Thames alone. (Alexander Turnbull Library)
(Inset) Now full of exhibits, the Thames School of Mines Museum began its life as a Methodist Sunday School.
The Golconda Tavern (left), a major landmark in the historic town of Coromandel, an hour's drive north of Thames.
(Above) Civic respect — Coromandel's council chambers. The war memorial statue, like hundreds of monuments all around the country, was carved from granite cut at the Old Granite Wharf (below) on the north-west Coromandel coast.

(Far left) Te Kouma, a tranquil anchorage on historic Coromandel Harbour, is also well known as a popular destination in the Royal New Zealand Yacht Squadron's annual Commodore's Cup Passage Series. (Left) Looking out to Little Barrier Island from Fantail Bay. The northern tip of the Coromandel Peninsula is one of New Zealand's most unspoiled areas. Colonies of sea birds, including shags (below) and gannets can be found all along the Coromandel coast.

As with Great Barrier Island, the Coromandel Peninsula was ruthlessly exploited from the earliest days of European settlement. But, again like Great Barrier, the Coromandel survived and remains serenely beautiful.
(Right) The solitude to be found on the shores of the Gulf is unequalled.
(Below) Holy Trinity Church, Thames, built entirely with Maori labour.

MER
CU
ON 700
RADAR
Caro Vita

Chapter 6

Sail City

(Pages 152-153) Well-wishers crowd Merit Cup *as she sets sail for Brazil on the fifth leg of the 1997-98 Whitbread round-the-world race (now Volvo Ocean Race).*
(Above) The late Sir Peter Blake returns to New Zealand with the Auld Mug. Blake was an icon of New Zealand sailing, knighted in 1991 and tragically killed 10 years later.
(Photo New Zealand Herald*)*
(Above right) Triumphant home-coming. The 'City of Sails' welcomes Peter Blake, Russell Coutts and Team New Zealand.
(Photo: New Zealand Herald*)*

Quest of Quests

In May 1995, the 'City of Sails' stopped working and started cheering. Aucklanders in their thousands thronged the streets to welcome home Team New Zealand with the America's Cup. The tumultuous scenes were to be repeated as the holy grail of yachting was paraded around the country. Over the previous 25 years, the nation's sailors had won the Whitbread round-the-world race, the Admiral's Cup, the World Match Racing Championships and various Olympic gold medals, but this was New Zealand yachting's crowning achievement. For only the second time in 144 years, the Auld Mug had been wrested from America.

According to Dennis Connor, 'The America's Cup is a game of life.' It's a game that began in 1851 when, to mark the first Universal Exhibition, a black schooner called *America*, representing the New York Yacht Club, was invited to race around the Isle of Wight. Against all odds, she beat the reputable British fleet of yachts. The prize, a solid silver Victorian urn, crossed the Atlantic to the United States. The British immediately challenged and the America's Cup, the world's most famous sailing race, was born.

The New York Yacht Club retained their grasp on the Cup for 132 years. Then, in 1983, *Australia II*, helmed by John Bertrand, snatched victory and the Auld Mug disappeared Down Under. Dennis Connor, the first American to lose the Cup, won it back at Fremantle in 1987. He then lost it again, this time to Team New Zealand and *Black Magic*, off the shores of San Diego in 1995.

Bringing the Auld Mug home to New Zealand took over 10 years of effort and four challenges. As Team New Zealand spokesperson Alan Sefton points out, the America's Cup is probably the hardest event in sport to win. Firstly, millions of dollars in sponsorship must be raised. Then, a large team of designers, boatbuilders, sail makers, sparmakers and crew must remain highly focused over a long period of time. 'A winning syndicate is like a Formula 1 racing car,' says Alan Sefton. 'You've got to stay tuned.' An apt analogy. In 1995, *Black Magic* blitzed *Young America* 5-0.

The 30th battle for the America's Cup was fought out on the Gulf in 1999-2000, with Italy's Prada Challenge winning the Louis Vuitton trophy and the right to challenge Team New Zealand for the America's Cup. The New Zealand team, skippered by Russell Coutts and Dean Barker once again demolished the challenge, winning 5-0 and earning New Zealand a place in the history books as the first non-American team to retain the Cup.

(Above) The regatta courses for the 30th defence of the America's Cup. The course for each individual race is set on the basis of the prevailing winds.

(Right) A familiar sight on the waters of Auckland's Hauraki Gulf in early 2000 was that of Team New Zealand's 'Black Boat' leading Italy's Prada Challenge, to eventually win the Auld Mug in convincing style.

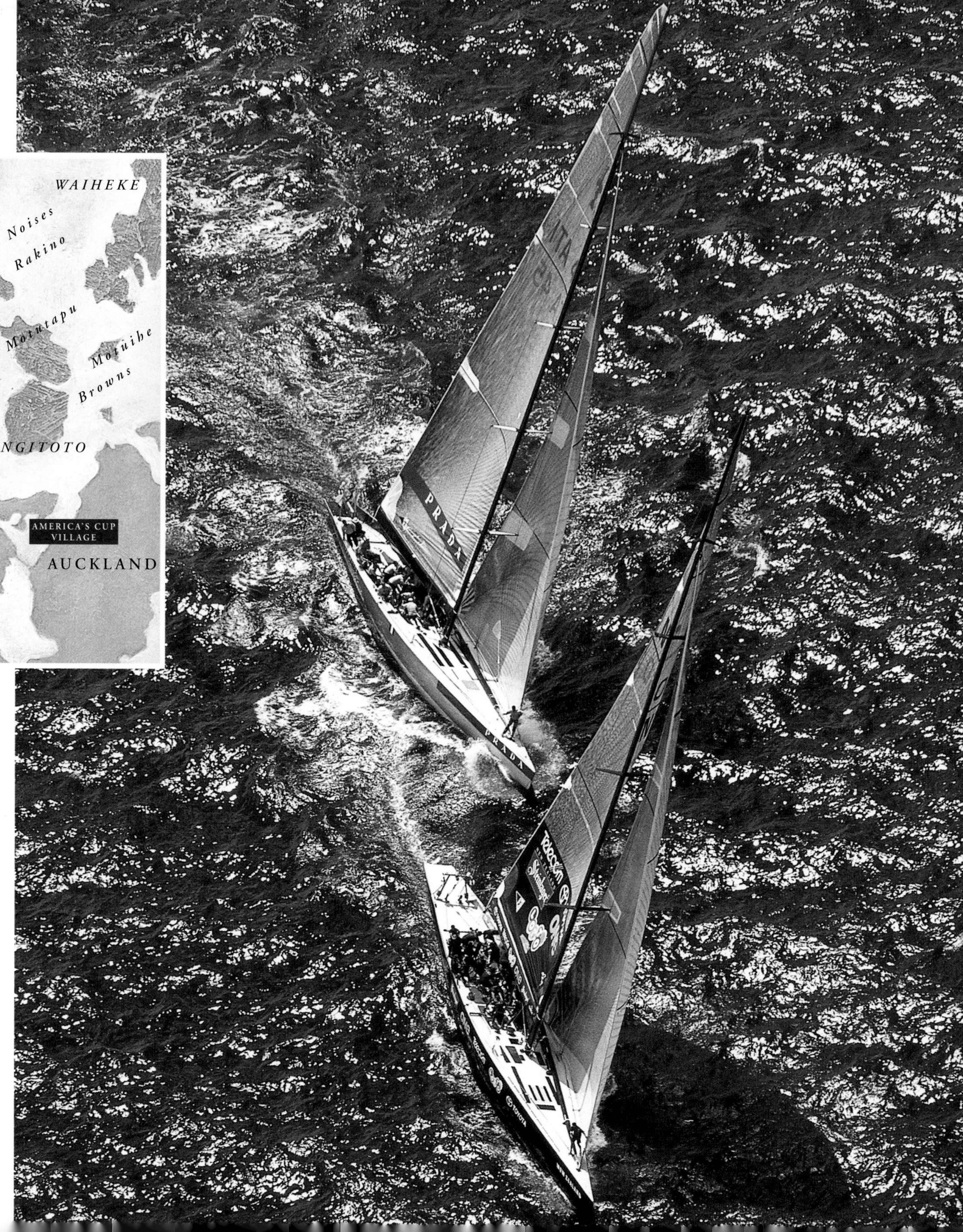

(Above) *Time out at Auckland's Viaduct Basin for the 1997-98 Whitbread (now Volvo Ocean Race) round-the-world fleet.*

Blue Water Marathon

Two decades ago, the first Whitbread yachts to sail down the Hauraki Gulf and into the Waitemata Harbour received a rousing welcome from Aucklanders. It was the second round-the-world race — Sydney was the stopover for the first in 1973-74 — and ever since, the 'City of Sails' has been a favourite port of call with Whitbread (now Volvo Ocean Race) crews.

Originally organised by Colonel Bill Whitbread of the famous English brewing family, and Admiral Otto Stein of the Royal Naval Sailing Association, the world's supreme ocean yacht race has changed dramatically over the years. The first boats mainly belonged to wealthy individuals and were manned by non-professional crews — skippers and navigators rarely knew where their rivals were and there was a weekly report to the nearest coast-guard. Today, the boats are purpose-built featuring state-of-the-art technology, and are often sponsored by multi-national corporations, the vast majority of the crews are professional sailors and media coverage of the event is extensive. A change of sponsorship and format see the now nine-leg race called the Volvo Ocean Race. Auckland hosts the boats at the end of leg three, approximately halfway through the 32000 namutical miles of the race.

MER

(Left) Merit Cup, *skippered by kiwi Grant Dalton, charges down the Hauraki Gulf on her way to the Orakei Wharf finish line and victory in the Sydney-Auckland leg of the 1997-98 Whitbread round-the-world race. (Photo:* New Zealand Herald*)*
(Far left inset) A section of the huge welcoming crowd gathers on North Head.
(Right insets) Merit Cup *powers up the Gulf past Browns Island and Bean Rock, and over the finish line.*
(Page 160) Team New Zealand return to the Viaduct Basin triumphant after their 5–0 victory over Italy's Prada Challenge in the 2000 America's Cup. Thousands of spectators gathered to welcome them back to base after one of the most thrilling contests to take place on the Hauraki Gulf.

LOADED HOG
TOYOTA
Telecom
NEW ZEALAND
TOYOTA
200
200